# TRIBUTE TO THE WESTERN REGION HYDRAULICS
## TOM HEAVYSIDE

The hydraulics were equally at home on freight as well as passenger services. 'Western' class No. 1069 *Western Vanguard* clatters towards Horton Road Junction, Gloucester, with a lengthy mixed freight for South Wales from the Swindon direction on 16 September 1975. 'Peak' class diesel-electric loco No. 45057 waits patiently in the loop by the connecting line from Barnwood Junction, on the Birmingham-Bristol line. The 'Western' was withdrawn the next month.

© Tom Heavyside, 2021
First published in the United Kingdom, 2021,
by Stenlake Publishing Ltd.
www.stenlake.co.uk
ISBN 978-1-84033-902-4

The publishers regret that they cannot supply copies of any pictures featured in this book.

Printed by P2D Books, 1 Newlands Rd, Westoning, Bedford MK45 5LD

## Acknowledgements

As with previous volumes in this series I am again indebted to Paul Abell, Peter Barber and Eddie Johnson for their ready help. Thank you gentlemen. I should also acknowledge the many unsung individuals who ensured a number of hydraulic locomotives were saved for posterity, and thus can continue to be admired at a number of heritage railways around Britain.

'Warship' No. 824 *Highflyer*, 'Hymek' No. 7038 and 'Western' No. 1011 *Western Thunderer* rest in the stabling sidings by Exeter St David's station on 13 October 1972. Officially the D prefix to the numbers on diesel locomotives had been dropped following the end of steam on normal services on British Rail in August 1968, the D usually being obliterated as on *Highflyer*. However, when new the 'Hymeks' were fitted with raised aluminium alpha/numeric digits, and while on some members of the class the D was chipped off and on others painted over, many as No. 7038 retained the prefix. As regards the 'Western' class which had cast number plates, in a style reminiscent of those attached by the Great Western Railway to its steam locomotives, the D remained in position.

# Introduction

When British Railways began to give serious consideration to the elimination of steam traction during the mid-1950s, the Western Region management decided that their main line diesel locomotives would use hydraulic transmission, while the other regions all concentrated their resources on diesel-electric power. However by the late 1960s, due to the reduction and changing patterns of freight traffic, and a need to standardise the locomotive stock, with far greater numbers of diesel-electrics in service inevitably the axe fell on the hydraulics. Thus their reign as the dominant power on the Western Region was relatively brief.

Initially, in November 1955, the Western Region signed a contract with the North British Locomotive Company of Glasgow for the supply of five Type 4 2,000hp locomotives utilising two German MAN engines, Voith transmission and a A1A-A1A wheel arrangement. Weighing 118 tons and with a maximum speed of 90 mph, they were despatched south from the maker's Queen's Park factory in a standard green livery numbered D600 to D604 between December 1957 and January 1959. They were distinguished by being named after Royal Navy warships and referred to as the 'Warship' class. Attached to Plymouth Laira depot, when new they were employed on such prestigious trains serving London Paddington as 'The Bristolian' and 'The Cornish Riviera Express', but following the arrival of the D800 series were soon relegated from front-line duties. They spent much of their latter days working west of Plymouth, before they were withdrawn in December 1967.

No. D800, the first of the Type 4 series based on the already proven Deutsche Bundesbahn V200 class, although considerably slimmed down compared to the German model due to the more restricted British loading gauge, was completed at Swindon Works in June 1958. It had two Maybach engines and Mekydro transmission, a B-B wheel arrangement with a top speed of 90 mph, but was considerably lighter than the D600 series at 79 tons. Swindon went on to build Nos. D801 to D832 and D866 to D870, the last being released to traffic in October 1961. No. D830 was experimentally fitted with Paxman engines. The North British Locomotive Company was responsible for Nos. D833 to D865, these like their predecessors from the same factory using two MAN engines and Voith transmission, which were added to stock between July 1960 and June 1962. While Nos. D800 to D802 could develop 2,000hp, the remainder had engines uprated to provide 2,200hp.

They were painted in a green livery with a horizontal white band along the body sides, and other than the first of class No. D800 *Sir Brian Robertson*, and No. D812 *Royal Naval Reserve 1859-1959*, they too were personalised with the names of British warships, from No. D803 *Albion* to No. D870 *Zulu* in alphabetical order, and also officially known as the 'Warship' class. From the summer of 1968 when BR introduced a numerical classification system, the Swindon examples became denoted as Class 42 and their North British counterparts Class 43. Earlier, from September 1965, beginning with No. D857 *Undaunted*, thirty-two were repainted maroon, but all except eight ended their days in BR standard blue.

Based for most of their lives at Newton Abbot and Plymouth Laira depots, the 'Warships' powered a variety of trains from the crack expresses to various freight trains, and could sometimes be seen as far north as Crewe. A far reaching development in January 1963 was the Western Region gaining control of the former Southern Region territory west of Wilton, near Salisbury, and from September 1964 the Exeter to London Waterloo services were added to the 'Warships' duties.

The withdrawal of the 'Warships' began when the slightly lower powered Nos. D800 to D802 were deleted from stock during the second half of 1968, four more being expunged from their ranks the next year. Forty-five were retired during 1971, leaving nineteen, all Swindon-built examples, on the books of BR at the start of 1972. The last five survivors were written off at the end of that year.

The next class of hydraulics to enter service were the Type 2 B-B (later Class 22) locos built by the North British Locomotive Company. Fitted with a MAN engine rated at 1,000hp and Voith transmission, the prototype No. D6300 was delivered to the Western Region in January 1959. Starting with No. D6306 the engines were uprated to 1,100hp, and utilising Clayton or Stones steam heating equipment rather than Spanner, at 68 tons these locos weighed 3 tons more than the original batch. The last of the class No. D6357 left Glasgow in November 1962. They were limited to a top speed of 75 mph. When new they were all painted green but later in life twenty-seven received a coat of blue.

The Type 2s were intended for secondary duties in the West of England, including pilot work over the steeply-inclined south Devon banks between Newton Abbot and Plymouth, but from the autumn of 1963 some were based at Old Oak Common (London), mainly for empty coaching stock movements in and out of Paddington. They soon became surplus to requirements and No. D6301 was taken out of service in December 1967, followed by a further twenty-six in 1968 and two in 1969. The rest of the class remained unscathed until 1971 when twenty-five had their engines shut down for the last time, the final four being withdrawn on 1 January 1972.

The need for locomotives in the Type 3 power range for mixed traffic duties was fulfilled by a consortium known as Beyer Peacock (Hymek) Ltd established by Beyer Peacock of Manchester, Bristol Siddeley Engines and J. Stone & Co. of Deptford, 101 being supplied from the former's Gorton factory in Manchester between May 1961 and February 1964. Numbered from D7000 and later classified 35, they had Maybach engines rated at 1,700hp and Mekydro transmission, a B-B wheel arrangement, turned the scales at 74 tons and were able to travel at 90 mph. The 'Hymeks', as they were commonly known, were delivered in a green livery and became a common sight almost throughout the Western Region. However their appearance started to change from November 1966 when Swindon began to apply an outer layer of blue paint during works visits, although thirteen ended their days in the original green.

In the event, like the rest of the hydraulic fleet, the majority of the 'Hymeks' were destined to enjoy but comparatively short existences. Nos. D7006 and D7081 were the earliest marked for the cutter's torch in September 1971, a further sixteen being condemned by the end of the year and no less than sixty-two during 1972. With the death throes seemingly in sight, BR organised a 'Hymek Swansong' railtour for Saturday 22 September 1973, Nos. D7001 and D7028 heading the special from Paddington to Hereford via the Severn Tunnel and return via Worcester. However, due to a shortage of power, six 'Hymeks' managed to linger on into 1975, the last four finally bowing out in March of that year.

Earlier, during the late 1950s, the Western Region realised the need for a more powerful Type 4 diesel locomotive, and Swindon began work on the design of what became the 'Western' class – later Class 52. Seventy-four of these handsome C-C locomotives took to the rails from December 1961 utilising two Maybach engines generating 2,700hp and Voith transmission, again with a top speed of 90 mph. They weighed 108 tons. Swindon Works completed Nos. D1000 to D1029, Nos. D1030 to D1073 being built at Crewe. All were in service by July 1964.

The doyen of the class left Swindon sporting an experimental desert sand livery, while No. D1001 was painted maroon. Nos. D1002 to D1004 and the first four from Crewe, Nos. D1035 to D1038, were painted green

Young enthusiasts take a rubbing of the cast number plate of No. D1023 *Western Fusilier* at York on 12 February 1977.

with No. D1015 appearing in golden ochre, before maroon was decided as the preferred colour. Eventually all were repainted blue. Their appearance was enhanced by cast number plates, similar in style to those traditionally attached to steam locomotives by the Great Western Railway. Horizontal nameplates, all prefixed 'Western' decorated the sides, the first No. D1000 appropriately named *Western Enterprise*.

During their lives the 'Westerns' were employed on passenger and a variety of freight trains, the class remaining intact until May 1973 when Nos. D1019 *Western Challenger* and D1032 *Western Marksman* were condemned. As the withdrawals continued apace during the ensuing years, more and more railway enthusiasts began to take a keen interest in seemingly their every movement, and the class gained almost a cult following. During 1976 and into 1977, as the end of the hydraulic era ebbed ever closer, the class was used on many specials, often taking them to unusual destinations, on a couple of occasions No. D1023 *Western Fusilier* even travelling as far north as York.

While the other hydraulic classes faded from view on BR with few taking much notice, in contrast the 'Westerns' departed in a blaze of glory. BR organised the final requiem to the class, the 'Western Tribute' which ran from Paddington to Swansea, then to Plymouth before returning to London on Saturday 26 February 1977. Nos. D1013 *Western Ranger* and D1023 *Western Fusilier* hauled the sell-out train, which in case of mishap was shadowed for much of the journey by Nos. D1010 *Western Campaigner* and D1048 *Western Lady*. Many thousands lined the route to pay homage to a much loved class, the event not going unnoticed by the national press.

Over two-and-a-half years elapsed after the first of the 'Westerns' had made their debut before the last of the hydraulic classes, a Type 1 (later Class 14) with a 650hp Paxman engine and Voith transmission, was released from Swindon Works in July 1964. Numbered D9500 this 0-6-0 turned the scales at 50 tons with speed limited to 40 mph. It was readily recognisable by a raised slightly off-centre cab, with the bonnets painted Brunswick green and the cab a lighter shade of Sherwood green. In total fifty-six were built, all at Swindon, the last No. D9555 in October 1965. Intended for shunting, branch line and short transfer freights, they were allocated to Old Oak Common (London), Bristol Bath Road, Worcester, Cardiff Canton and Landore (Swansea) depots.

However, suitable work for the Type 1s was rapidly diminishing, and it was not long before they became redundant, with Nos. D9522 and D9531 being cast aside as early as December 1967; in an effort to find them suitable work thirty-three were transferred to Hull Dairycoates shed on the Eastern Region, only for them to be withdrawn *en masse* in April 1968. Eleven of those that remained loyal to the Western Region survived into 1969, these being stopped in March and April of that year. All fifty-six had scandalously short BR careers, No. D9554 being on the books for just two-and-a-half years, while the longest to remain in capital stock, Nos. D9500 and D9502, both built July 1964, were among the last to be culled in April 1969 when less than five years old! However only eight were to face an oxyacetylene torch more or less immediately after withdrawal, the remainder (some after resting in various scrapyards for a short time) being snapped-up by industrial concerns, the National Coal Board and the British Steel Corporation being the principal beneficiaries.

When the hydraulics roamed the Western Region interest in diesel locomotives was in its infancy. Consequently, sadly, no representatives of the North British-built engines have survived into the preservation era, this despite Nos. D600 *Active* and D601 *Ark Royal* being sold for scrap to Woodham Bros. of Barry, South Wales, from where over 200 steam locomotives were rescued, and where the latter was not dismembered until June 1980. Fortunately the products of Swindon, Crewe and Beyer Peacock fared better, two 'Warships', four 'Hymeks' and seven 'Westerns' escaping the clutches of the scrap-metal merchants. Ironically, no less than nineteen of the Class 14s, due to their second careers with industrial concerns, have been preserved. These now much-respected machines can be seen at various heritage railways and museums around Britain.

As a Lancashire man born and bred, during the late 1950s and 1960s the hydraulics were not part of my normal diet. However, on numerous visits south during those years, albeit principally in search of steam, I was fortunate to witness some of their early exploits, although with hindsight not pointing my camera in their direction often enough. I gave them much more attention after the end of steam on BR in 1968, before their days were so cruelly cut short by the decision to standardise on diesel-electric locomotives. This volume is but a small tribute to their all-too-brief years on British Railways.

Tom Heavyside
Bolton, Lancashire
January 2021

A few days after release from Swindon Works, the next-to-last of the 365 diesel-hydraulic locomotives built for the Western Region, Type 1 No. D9554, unexpectantly finds itself in illustrious company at the adjacent shed on Saturday 9 October 1965. The legendary 1923-built LNER Pacific No. 4472 *Flying Scotsman*, then in private ownership, had had to be relieved at Swindon while hauling the *Railway Magazine* sponsored 'Welsh Mystery Flyer' excursion from London Paddington to Cardiff. It had been towed to the shed by 'Hymek' No. D7055. While No. D9554 had a life of only two-and-a-half years on BR before being sold to Stewarts & Lloyds Minerals Ltd (later British Steel Corporation), for service at their Corby Ironstone Quarries, Northamptonshire (where it was scrapped in August 1982), *Flying Scotsman*, now part of the National Collection, can still be seen out and about on the main lines and at heritage railways.

No. D9513 as NCB No. 38 shunts three wagons and a brake van at Embsay, near Skipton, on the Embsay Steam Railway on 14 July 1991. Note the miniature ground signals by the foot crossing. Life on BR for No. D9513 lasted from October 1964 until March 1968. It subsequently worked at National Coal Board Opencast Executive sites at Crigglestone near Wakefield, and Astley near Leeds, before moving to Northumberland in January 1974 where it was employed on colliery railways at Ashington, Backworth and Burradon. Preservation beckoned in October 1987 and the loco can still be seen at Embsay, now known as the Embsay & Bolton Abbey Steam Railway.

In their early days the hydraulics had to share depot space with steam locomotives – a far from ideal arrangement. Here at Plymouth Laira shed on 24 May 1960, four months old Type 2 No. D6314 basks in the afternoon sunshine alongside elder statesman 'King' class 4-6-0 No. 6002 *King William IV*. Both locomotives were allocated to Laira. Note No. D6314 has its front end communication doors open, as when working in multiple, and no yellow warning panel as was the norm when diesel locos were first introduced on BR. The influx of the hydraulics meant the reign of the 'Kings' as the supreme form of motive power on the Western Region was fast drawing to a close, all thirty members of the class being withdrawn during 1962.

*Opposite*: During the early 1960s diesel and steam combinations were not uncommon, especially between Plymouth and Newton Abbot where steep climbs in both directions to summits at Wrangaton and Dainton have to be surmounted. Here 'Warship' No. D807 *Caradoc* receives a helping hand from 'Castle' class 4-6-0 No. 5096 *Bridgwater Castle* as they leave Plymouth (formerly North Road station) with a train from Penzance, with portions destined for both Manchester and Glasgow on 30 May 1961. Note the Royal Mail Travelling Post Office coach behind *Caradoc*, equipped for collecting and depositing mail at speed from lineside apparatus. The duo are about to pass a second 'Warship' approaching from the east. Coincidentally 'Grange' class 4-6-0 No. 6873 *Caradoc Grange*, allocated to Laira along with No. D807 (No. 5096 was a Taunton engine), was also seen that day. In total during the day, including a visit to Laira shed, the author noted 120 locomotives, seventy-six steam and forty-four diesels. Comparable figures for the visit to Plymouth the previous year on 24 May 1960 were eighty-seven steam and twenty-nine diesels.

Type 2 No. D6325, in its early green livery, nears Par station from the west with a rather mixed rake of stock for the permanent-way department on Sunday 28 May 1961. The position of the one open white disc on the loco, indicates this to be an inspection train (also used for ballast or freight) that may need to stop between signal boxes. Allocated to Laira from new in June 1960, No. D6325 was withdrawn from Bristol Bath Road depot in October 1968. It was later bought as scrap by John Cashmore Ltd of Newport.

*Opposite*: On the same day as the previous picture, two porters at Par prepare to greet passengers alighting from the 'Cornish Riviera Express' ex-London Paddington behind 'Warship' No. D807 *Caradoc*. Travellers bound for Newquay will need to cross the line to the outer face of the platform on the left. The connections to the branch off the main line can be seen just beyond the road bridge. Note the metal frame attached to the nose of *Caradoc*, used for train identification purposes (see page 15), similar to those affixed to steam locomotives on the Western Region. On this occasion there is no outward recognition of this prestigious service, either by train number or headboard, although the position of the two white discs confirm this to be an express passenger service. The loco entered traffic in June 1959 and when withdrawn in September 1972 had a recorded mileage of 1,317,000, the highest accumulated by a 'Warship'.

With the town's gas works behind, Type 2 No. D6307 has just arrived at Newquay, having travelled the 20¾-mile branch from Par, conveying goods vans as well as passenger accommodation, again on 28 May 1961. The second man is attending to the hinged upper train disc, the three discs at a lower level being in the closed position. Prior to 4 February 1963 this popular Cornish coastal resort was also served by a branch from Chacewater, west of Truro, on the main line to Penzance. The author had returned to Newquay after a round trip to Truro and Par, ten Type 2s, seven 'Warships' and six diesel-electric shunters being logged, and although thirty-six steam locomotives were also noted, this indicated the growing importance of diesel traction. The next year, 1962, the three ex-GWR sheds in the Duchy, St Blazey (Par), Truro and Penzance, were all closed to steam.

Back at Plymouth, when the station was in the midst of a six-year reconstruction scheme that had started in 1956, 'Warship' No. D815 *Druid* waits to leave with the Penzance-Paddington 'Royal Duchy' service on 24 May 1960. The relevant headboard has not been placed on the locomotive, but in common with the other titled trains on the Western Region at this time, the carriages are resplendent in the revived former chocolate and cream colours of the erstwhile Great Western Railway. From the previous December when No. D813 *Diadem* was released from Swindon Works, the steam age metal frames used to display the train reporting number (see page 11) were abandoned in favour of four-character illuminated hand-wound blinds as seen here, a refinement that meant the previous headcode discs were no longer needed. For some reason on this occasion *Druid* is only showing the letter A to indicate the train is bound for London.

The 'Torbay Express' started its 208-mile journey to Paddington at Kingswear by the River Dart, where nine-months old No. D825 *Intrepid* receives a wash prior to departure time on 31 May 1961. The properties to be seen above the carriage on the right are in Dartmouth on the opposite bank of the river. Note the distinctive cream-backed, brown-lettered headboard incorporating the arms and motto *Auxilio Divino* (by Divine Aid) of Devon County Council, on the front of the loco. The 'Warships' first took charge of the 'Torbay Express' on Monday 27 July 1959 when No. D807 *Caradoc* hauled the train from Kingswear, and No. D808 *Centaur* headed the corresponding down service from Paddington. No. D825 was first withdrawn in January 1972, only to be reinstated in May 1972 along with Nos. D814 *Dragon* and D829 *Magpie* due to a shortage of motive power, the trio then running without nameplates until final withdrawal in August 1972. The rails to Kingswear were originally laid by the Dartmouth & Torbay Railway in 1864: the ambitious plan to bridge the Dart never materalised, although a station was built in Dartmouth and a connecting ferry service provided. Today Kingswear is the terminus of the Dartmouth Steam Railway, the line from Paignton having been operated as a heritage railway since 1973.

*Opposite*: The next day the 'Torbay Express' approaches Torquay behind the doyen of the Swindon-built 'Warships' No. D800 *Sir Brian Robertson* (Chairman of the British Transport Commission 1953-1961). Note the far less ornate black-backed headboard and the train reporting number A68 utilising the metal frame holder, the class of train – express passenger – denoted by the two white discs. Awaiting its arrival is Hawksworth-designed 0-6-0PT No. 9440, ready to buffer up to the rear to provide banking assistance up the 1¾-mile climb to Torre, initially at 1-in-55 before easing to 1-in-73. The pannier tank had been based at Newton Abbot since supplied new to BR by Robert Stephenson & Hawthorns of Newcastle-upon-Tyne in February 1951. It was withdrawn in July 1963 after spending its last year at Old Oak Common (London) shed. The 'Warship' had a slightly shorter lifespan of ten years four months, being cast aside in October 1968 after travelling some 923,800 miles. It was subsequently scrapped by John Cashmore Ltd of Newport in July of the following year, the only D800 series loco to go to its grave still wearing the original green paint scheme.

A rather unkempt Type 2 No. D6337, with a small yellow warning panel, nears Kingskerswell with a local train down the Kingswear branch, having left the Newton Abbot-Plymouth main line at Aller Junction a couple of minutes earlier on 9 April 1966. Starting with No. D6334, as seen here North British equipped the locomotives with split four-character roller blinds in place of the previous discs (see pages 10 and 12). No. D6337 was retained in capital stock until October 1971, when it was withdrawn from Plymouth Laira depot, being disposed of the following May at Swindon Works, the last resting place of thirty of the Type 2s.

*Right*: The classic view of Newton Abbot observed earlier on the same day as the picture opposite, as 'Hymek' No. D7029 passes along the down through road with three coaches in tow. In the right background is the site of the former steam shed, and while the bulk of the steam allocation had been transferred away by mid-1962, a few lingered on until January 1963. Four years earlier in March 1959 seventy-four steam locos had had an 83A (Newton Abbot) shed plate attached to their smokebox doors, including no less than thirty with a 4-6-0 wheel arrangement, fourteen 'Castles', twelve 'Halls' and four 'Granges'. A 'Warship' has just emerged into daylight from the much more sanitised facilities demanded by the diesels. The 1962-built 'Hymek' is one of four that still survive and now resides on the Severn Valley Railway at Kidderminster.

*Left*: Moving across the road from where the previous photograph was taken, Newton Abbot-allocated 'Warship' No. D835 *Pegasus* has permission to proceed towards the station with a three-coach train off the Kingswear branch. Note the lower quadrant signals and the diesel shunter beyond on station pilot duty. In July 1960 this became the third D800 series 'Warship' despatched south by the North British Locomotive Company from their Glasgow workshops. With the addition of small yellow warning panels in June 1962 it retained its original green livery until March 1968, when it was outshopped after attention at Swindon Works in blue with full yellow ends. It was deemed surplus to requirements at Bristol Bath Road depot in October 1971, having clocked 785,000 miles.

The photographs on these two pages were also taken on 9 April 1966. Here maroon-liveried 'Western' class No. D1050 *Western Ruler*, complete with a small yellow warning panel, enters Newton Abbot from the west. On the right is Newton Abbot West signal box, opened in 1925 in conjunction with the remodelling of the station, work which was finished two years later when six roads became available through the site. The box housed a 153-lever frame. It closed in May 1987 when its functions were taken over by the new Exeter power box, the station layout being rationalised at the same time. This meant the end for the fine GWR signals seen in this and the previous two pictures. On the left is the former wagon works which closed in 1972. No. D1050 was released from Crewe Works in January 1963 and was first allocated to Cardiff Canton depot. It remained in service until April 1975 when it was withdrawn from Plymouth Laira, being disposed of at Swindon the following April.

'Hymek' No. D7084 rounds the curve past the 206-lever Newton Abbot East signal box with a train from the Exeter direction. Diverging to the left are the former lines to Moretonhamstead and the Teign Valley route to Exeter. In common with many other hydraulics No. D7084 had an active life of less than ten years, being accepted by the Western Region from Beyer Peacock in June 1963 and withdrawn in October 1972. It was dismantled at Swindon the next month.

With Newton Abbot Power Station overlooking the scene, a rather scruffy looking No. D859 *Vanquisher*, which came south from the North British Locomotive Works in Glasgow in January 1962, has received authorisation to continue on its way towards Exeter from East signal box with a rake of empty 16-ton mineral wagons. The leading wagon, No. B135401, was constructed by BR at Derby in 1953, while the second wagon, No. B211030, emerged from the Pressed Steel Company's factory at Linwood in Scotland at about the same period. *Vanquisher* was the first 'Warship' delivered with small yellow warning panels. It had a life of nine years and two months before being retired in March 1971.

Having followed the east bank of the River Teign from Newton Abbot, in the then standard green livery with small yellow warning panels, No. D860 *Victorious* rounds the tight curve onto the sea wall at Teignmouth, as a cool breeze blowing off the English Channel keeps the temperature down on 30 May 1962. The train will hug the coast as far as Dawlish Warren, before turning inland alongside the River Exe towards Exeter, undoubtedly one of the most scenic railways in England. From February 1968 until its withdrawal in March 1971, *Victorious* was to be seen with blue body sides and full yellow ends. Its days finally ended when cut-up at Swindon in December 1971.

*Opposite*: 'Warship' No. D822 *Hercules*, with small yellow warning panels that had been applied the previous January, calls at Exeter St David's with the Paddington to Kingswear 'Torbay Express' on 28 May 1962. By this time chocolate and cream coloured carriage stock no longer held sway. From 1964 the 'Warships' would also set out for the capital from the west end of St David's, but by the stiff climb to Exeter Central and on to Salisbury and London Waterloo. The lower quadrant signals were activated from Exeter West signal box, which – following decommissioning in November 1986 – was dismantled and re-erected at what is now the Crewe Heritage Centre. *Hercules* was taken out of service in October 1971, one of twenty-nine 'Warships' (all Swindon-built examples) to complete over one million miles.

'Warship' No. D824 *Highflyer* heads away from Exeter towards Taunton at Cowley Bridge Junction on 13 October 1964. The junction was the former London & South Western Railway gateway to Plymouth, Barnstaple, Ilfracombe, Bude and Padstow, the L&SWR enjoying running powers over the ex-GWR main line between here and Exeter St David's. The slightly lower right-hand arms on the bracket signals indicated when a train was destined for one of these routes. Today it is normally only possible to reach Barnstaple by former L&SWR tracks, except on summer Sundays when trains run along the former Plymouth line from Coleford Junction as far as Okehampton in connection with the heritage Dartmoor Railway based at the latter. When deleted from capital stock in December 1972, *Highflyer* was one of the last in service, having recorded 1,077,000 miles during its 12 years of service.

*Opposite above*: Maroon-liveried 'Warship' No. D805 *Benbow* surges up the steep climb from Wellington at Beam Bridge towards Whiteball summit, between Taunton and Exeter on 28 May 1966. The bridge on the left carries the main A38 road, while a learner driver carefully negotiates the minor road beneath the railway. After taking to the rails at Swindon Works in May 1959, when withdrawn in October 1972 *Benbow* had accumulated some 1,281,000 miles, the joint third highest recorded by the 'Warships'. No. D806 *Cambrian* was the other loco to travel a similar distance.

*Opposite below*: As in steam days freight trains often required banking assistance from Wellington to the summit at Whiteball. On the same day as the picture above at Beam Bridge, but viewed from the A38 overbridge, 'Hymek' No. D7084 provides the necessary rear end power up a 1-in-86 section, the leading engine 'Warship' No. D811 *Daring* being virtually out of sight. The consist will soon enter Whiteball Tunnel and on exiting will have passed from Somerset into Devon.

23

The second 'Warship' constructed at Swindon in November 1958, Laira-allocated No. D801 *Vanguard*, slows alongside platform 5 at Taunton with a Paddington-Penzance service on 5 September 1960. 'Hall' class 4-6-0 No. 5935 *Norton Hall*, built at Swindon twenty-five years earlier in July 1933 and based at Westbury shed (82D), stands at platform 1 ready to follow with a stopping service to Exeter. The 'Hall' was marked for scrap in May 1962, but even so enjoyed a lifespan almost three times that of *Vanguard*, which had the dubious distinction of being the first of its class withdrawn in August 1968.

No. D848 *Sultan* approaches Durston, soon after passing Cogload Junction, east of Taunton, with the northbound 'Devonian' *en route* from Paignton to Bradford in June 1967. Note the last two digits of the headcode have been amended to 37 from that which denoted this service in June 1961 (see inside front cover). The down line is at a higher level in order to bridge the Westbury lines, which avoids conflicting movements at this busy junction. Having been painted maroon in June 1966, the locomotive was destined not to live quite so long as its namesake battleship that took to the waters in 1870, and while transferred to harbour service in 1906 was not broken up until 1946. No. D848 was one of three 2,200hp machines, along with Nos. D830 *Majestic* and D863 *Warrior*, condemned in March 1969, following the demise of the three lower rated Nos. D800-D802 the previous year. When withdrawn, records show *Sultan* had covered 537,000 miles in revenue-earning service during the previous eight years: only the unique Paxman engine-fitted loco No. D830 had a lower return of 447,000 miles. No. D848 was one of eight 'Warships' not to receive a coat of blue paint.

Observed from by the entrance to Bristol Bath Road depot, 'Western' class No. D1047 *Western Lord* leaves Temple Meads station with the empty stock of a train from Paddington on 16 October 1965. On shed awaiting their next duties are (nearest the camera) 'Western' class No. D1009 *Western Invader* with a 'Hymek' almost obscured behind. Above the duo can be seen a 'Warship' and a Brush Type 4 (later Class 47), and behind the water tower a second 'Hymek' with a 'Peak' class just nosing into the picture on the right. Steam had been banished from the shed in September 1960, when the locos still required locally were dispersed to the other two sheds within the city at St Philip's Marsh and Barrow Road, Bath Road then concentrating its resources on diesel traction. In November 1965, shortly after the last of the hydraulic fleet had taken to the rails, eleven 'Westerns', sixty 'Hymeks' and fifteen Type 1s were based here.

*Opposite*: An unexpected outing for No. D7089 on 9 October 1965. The 'Hymek' had just been commandeered at Swindon to take forward the *Railway Magazine* 'Welsh Mystery Flyer' excursion to Cardiff, ex-LNER Pacific No. 4472 *Flying Scotsman* having had to be detached here on arrival from Paddington (see page 6), hence the interested onlookers and the many heads looking out from the carriage windows. Note, too, the lower quadrant signals. No. D7089 was retained on the active list until May 1973, but then after lying dormant until November 1974 was despatched to Laira and later transferred to Departmental Stock for carriage heating purposes. It was allocated departmental No. TDB968005 although this was never carried. Declared redundant for a second time in autumn 1975 it was bought by the scrap merchant T.J.Thomson of Stockton-on-Tees, where its days finally ended the following spring. It was the only 'Hymek' to be dragged so far north to its deathbed, and while a further ten were sold to private dealers, the remainder other than those purchased by preservationists, met their fate at Swindon Works.

In January 1963 administration of the former London & South Western Railway lines west of Wilton, near Salisbury, was transferred from the Southern to the Western Region. The next year in September the 'Warships' replaced steam on the Exeter to London Waterloo services. Here No. D818 *Glory* pulls away from Basingstoke with a ten-coach train from Exeter St David's bound for the capital on 26 May 1966. Under construction rising above the second and third carriages, is the electric operated signal box commissioned on 20 November 1966, which rendered many traditional boxes in the area, including Basingstoke A box (seen opposite), redundant. The signal gantry in the middle distance is at the east end of the station. The line to Reading is hidden from view within the cutting on the right.

Coming in the opposite direction on the same day, nearing Basingstoke with eight coaches forming a Waterloo to Exeter service is No. D819 *Goliath*. At the time the route was being prepared for electric units to take over the Waterloo to Bournemouth services the following year, and the conductor rail can be seen in position by some of the tracks. The name *Goliath* had earlier been attached to London & North Western Railway 'Jubilee' class 4-4-0 No. 1927 in honour of a battleship launched in 1898. No. D819 was one of those that succumbed to the mass withdrawal of 'Warships' in October 1971 (see page 34), having accumulated 1,125,000 miles on the clock.

Nineteen miles into its journey to Exeter with the 17.00 from Waterloo, No. D813 *Diadem* dashes through Weybridge along the down fast line, coinciding with a patch of sunlight on 12 May 1967. Behind the train can be seen the station platforms serving only the outer slow lines, together with the signal box on the right-hand side. The sub-station in the right foreground provided current to the electrified third rails. Branching off to the left is the line to Chertsey. In earlier times the name bestowed on No. D813 was conferred in 1901 on LNWR 'Alfred the Great' class 4-4-0 No. 1946, the name of a cruiser built in 1896. A second cruiser launched in August 1942 was also identified by the name. No. D813 had been painted maroon in November 1965 but spent its last years from July 1967 in blue. Mileage totalled 1,099,000 when withdrawn in January 1972.

With only 1½-miles remaining of its 172½-mile journey, No. D818 *Glory* rumbles through Vauxhall with an Exeter-Waterloo service on 7 October 1965. The station had eight platforms. Looking in the opposite direction, the author could see 'Big Ben' and the Palace of Westminster on the north bank of the Thames. No. D818 wore an outer coat of green paint until replaced by blue in June 1967. It was among the last nine 'Warships' to continue in service until the final months of 1972, being one of four withdrawn that November with the remainder being shut down for the last time the next month.

A panorama of Bristol Bath Road depot viewed from Temple Meads station with 'Hymek' No. 7047 nearest the camera on 14 October 1971. Early in 1967 the 'Hymek' had been repainted blue with off-white window frames and small yellow warning panels. While the D prefix to identify diesel locomotives was officially abandoned in August 1968 when BR rid itself of steam locomotives, as seen here the raised D remained *in situ* on many of the 'Hymeks'. To the left of No. 7047 are a couple of 'Warships' and representatives of classes 25 and 31, along with a Metro-Cammell Blue Pullman set. Clearly seen, even at this distance, are the all over yellow front ends of the 'Warships', which had replaced the small warning panels depicted in the earlier pictures. No. 7047 had only three months of life left before it was withdrawn in January 1972. It was subsequently disposed of at Swindon in August of the same year.

On the same day No. 7038 ambles through Bristol Temple Meads eastbound, under the splendid arched trainshed, with a mixed freight. Like its sister opposite, it too was in blue but with full yellow ends. It first saw the light of day at Beyer Peacock's Manchester factory in June 1962 and remained in service until July 1972. Its days also ended at Swindon, the oxyacetylene torches being directed towards its carcass in June 1973.

The sad sight of four recently withdrawn 'Warships' at Newton Abbot awaiting a tow to Swindon for scrapping on 15 October 1971. From left to right are Nos. 808 *Centaur*, 844 *Spartan*, 858 *Valorous* and 834 *Pathfinder*, the nameplates having already been removed. Note the BR double arrow symbol introduced during the mid-1960s on the body sides, which had gradually replaced the former lion holding wheel emblem. The stone building behind the 'Warships' was originally part of Newton Abbot locomotive works, the last steam loco to receive attention, Churchward 2-6-2T No. 4566, leaving with due ceremony on 15 July 1960. The buildings were then converted to a diesel maintenance depot – the fuelling facility can be seen on the right. Listed on its books in November 1965 were thirty-eight 'Warships' and seventeen Type 2s. The four 'Warships' depicted here were among the twenty-nine taken out of service that month, including the last sixteen North British examples. The depot closed its doors in October 1981.

*Opposite*: By the autumn of 1971 the 'Westerns' were the only hydraulic class still intact, the first withdrawals not occurring until May 1973. Here, blue-liveried No. 1056 *Western Sultan* departs from Plymouth with the leading seven coaches of the 'Cornishman' bound for Penzance on 13 October 1972. The train had started at Bradford; the rear portion, having been detached, awaits removal by the station pilot. The decorative headboards once carried by titled trains such as this were regrettably a thing of the past. BR staff occupied the large office block towering over *Western Sultan*. Note the loco has an all over yellow front end, first applied to No. 1048 *Western Lady* in place of the smaller warning panels (see pages 18 and 26) in November 1966. In the event it would be early 1972 before all the class had a similar outward appearance. No. 1056 remained in service until the end of 1976, and in company with the sixty-seven members of the class that were not fortunate to have a life beyond their days on BR, met its end at Swindon Works.

No. 1021 *Western Cavalier* propels a set of carriages from the sidings at Laira, on the eastern outskirts of Plymouth, before heading towards Plymouth station to form a service train on 24 June 1975. The tidal waters of the River Plym are on the right, while there is evidence of some road improvement schemes on the left. Laira diesel depot, opened in March 1962, was in sight behind the author. In November 1965 the depot had responsibility for 101 hydraulic locomotives, consisting of the five original North British and thirty-three of the D800 series 'Warships', twenty-five 'Westerns', fifteen 'Hymeks' and twenty-three Type 2s. From October 1971 all seventy-four members of the 'Western' class were based at the depot, and all ended their days there. No. 1021 was designated for scrap in August 1976, although it was not until March 1979 that it faced the cutter's torch at Swindon.

On the next day to the picture opposite, No. 1049 *Western Monarch* calls at Torquay with the 11.55 Paignton-Paddington service. On weekdays the semaphore signal on the right was usually to be seen in the lowered off position, the signal box near the far end of the down platform, which can just be discerned beyond the footbridge, normally being switched-out except on summer Saturdays, as it had been since October 1968. It closed permanently in November 1984. There is no sign of the former centre road through the station (see inside front cover). No. 1049 was delivered from Crewe Works to the Western Region in December 1962. It was taken out of revenue-earning service in April 1976.

Blue-liveried No. 1009 *Western Invader* guides an express service non-stop along the up through road at Newton Abbot *en route* to Paddington on 19 November 1973. The leading vehicle is a general utility van. Outshopped from Swindon Works painted maroon in September 1962, No. 1009 first operated from Old Oak Common depot in London. It was withdrawn in November 1976.

In pleasant Devon autumn sunshine, No. 1005 *Western Venturer* hurries by the down loop line at Whiteball, between Taunton and Exeter, with the 12.30 from Paddington to Paignton on 21 November 1973. The county boundary between Somerset and Devon lies within the 1,088-yards-long Whiteball Tunnel, the western mouth of which can be seen from the overbridge. No. 1005 was the second 'Western' to emerge new from Swindon in maroon livery in June 1962, after No. 1001 *Western Pathfinder* the previous February (Nos. 1002-1004 first appeared in green). It remained on the BR inventory until November 1976.

*Left*: After nightfall on the same day as the picture overleaf, No. 1059 *Western Empire* pauses at Taunton with the 15.30 Paddington to Penzance service. In March 1967 the centre island platform was closed to passengers, the adjacent tracks then being the preserve of non-stopping trains. It was reopened in May 2000. After completion at Crewe Works in April 1963 No. 1059 was first allocated to Cardiff Canton depot. It was retired from Plymouth Laira in October 1975 and subsequently dismembered at Swindon in July of the following year.

*Right*: No. 1034 *Western Dragoon* approaches Taunton from the west on its way to Paddington on 11 October 1971. This Crewe-built loco entered service in April 1964, the erection of Nos. 1030 to 1034 having been transferred from Swindon to Crewe. It was withdrawn in October 1975 after an active life of eleven-and-a-half years, but was then used for a short time for carriage heating purposes at Laira, before removal to Swindon for cutting up, which eventually took place in February 1979

No. 1042 *Western Princess* speeds by the former halt at Creech St Michael, two-and-a-half miles east of Taunton, on its way to Paddington with a train from the West of England on 11 October 1971. The halt closed on 5 October 1964, but a couple of small buildings on what remains of the up platform were retained for use by the permanent-way gangs. There is no trace of the former down platform. The tall chimney on the left belonged to Creech Paper Mills. The loco entered service from Crewe Works in October 1962 and was retained in capital stock until July 1973, an active life of ten years and nine months, eight 'Westerns' having even shorter existences, No. 1032 *Western Marksman* being retained by the operating authorities for only nine years and five months. No. 1042 was disposed of at Swindon in May 1974.

No. 1021 *Western Cavalier* rounds the tight curve from Heywood Road Junction so as to make the scheduled stop at Westbury with the 14.30 Paddington to Paignton service on 17 September 1975. Express trains to the West Country follow the 1933-completed Westbury cut-off line between Heywood Road and Fairwood junctions. The tracks veering to the left lead to Bath and Chippenham on the Bristol to Swindon route.

*Opposite*: On the same day No. 1043 *Western Duke* departs from Westbury with a service from the West Country to Paddington. Class 47 No. 47505 is ready to leave the adjacent platform at the head of the 16.20 from Weymouth to Cardiff. Westbury North signal box closed in May 1984 when the new Westbury power box, built on the vacant land to be seen to the right of *Western Cavalier* in the picture above, assumed control of the area. No. 1043 left Crewe Works new in October 1962 and was retained in capital stock until April 1976.

*Opposite*: Two photographs at the approach to Sonning cutting, east of Reading, on 3 October 1975. *Above*: No. 1041 *Western Prince* has charge of a long link of empty stone wagons destined for one of the Somerset quarries west of Westbury. Preservation beckoned after withdrawal in February 1977, the loco now being located on the East Lancashire Railway at Bury. *Below*: No. 1022 *Western Sentinel* follows a short time later on the fast line with the 14.05 from Paddington to Birmingham. Parked a few yards away, facing the same direction, is the author's red-liveried Renault car.

*Right*: A study in front ends under the magnificent arches at Paddington the previous day. No. 1015 *Western Champion* rests at the buffer stops alongside Class 47 No. 47013, after hauling a charter from Cornwall for pupils of Liskeard Grammar School. No. 1015 was initially turned out from Swindon in January 1963 sporting an experimental golden ochre livery. This was later changed to maroon before standard blue was applied. During the first year of its life it had two spells totalling eight months attached to Old Oak Common depot (three miles west of Paddington), one of twenty-three 'Westerns' based there in November 1963. By November 1965 the 'Western' allocation had been reduced to five, the depot then also having charge of eighteen Type 2s and twenty-six 'Hymeks'. No. 1015's BR days drew to a close in December 1976, but happily it was then rescued for preservation and from time to time has continued to be seen on main line excursions. Today it keeps company with fellow surviving 'Westerns' Nos. 1013 *Western Ranger* and 1062 *Western Courier* on the Severn Valley Railway at Kidderminster. No. 47013 (originally numbered D1540) was not so fortunate, being deleted from the books of BR in February 1987 and subsequently scrapped.

No. 1058 *Western Nobleman* pauses at Bridgend with a Paddington-Swansea service on 25 March 1976. By this time train reporting numbers had ceased to be displayed and opportunity has been taken to wind the blinds to confirm the loco's identity, much appreciated by lineside observers when trains were passing at speed. The banner repeater signal above the covered footbridge indicates the next signal on the up line, unsighted from this position, has been pulled off for an approaching train. In the car park a Ford Escort saloon nearest the camera, and a Singer Gazelle are the most prominent. While the loco had been based at Laira depot since April 1966, after its initial allocation in March 1963 to Bristol Bath Road it had moved to Cardiff Canton in June 1963 and then to Landore (Swansea) in March 1964, and thus in the mid-1960s was a familiar loco in the area. In fact in June 1964 Canton had twenty-five 'Westerns' on its books along with twenty 'Hymeks', while a further fifteen 'Westerns' were based at Landore. 'Nobleman' was one of seven 'Westerns' that remained in service at the start of 1977, but along with No. 1022 *Western Sentinel* was deleted from capital stock that January. No. 1058, together with No. 1028 *Western Hussar,* were the last two to be cut-up at Swindon in June 1979.

*Left*: On a rather dull day No. 1008 *Western Harrier* prepares to stop at Neath with a down parcels service, before continuing to Swansea on 26 May 1973. The locomotive was one of eleven culled during 1974, its eleven-year career starting and ending attached to Laira depot, although like its sister opposite it too spent time in South Wales in the mid-1960s. For the record, in November 1965 thirty-one 'Westerns' and thirteen of the Paxman-engined Type 1s were allocated to Landore, the only hydraulics then at Canton being twenty-six Type 1s.

*Right*: The sad sight of the first two main line hydraulic locos built for BR, Nos. D600 *Active* and D601 *Ark Royal* (nearest the camera), awaiting their fate in the famed Woodham Bros scrapyard at Barry, South Wales, on 22 October 1968. The five original North British 'Warships' were all written off in December 1967 and sold for scrap the following July, the other three, Nos. D602-D604, being purchased by John Cashmore of Newport, where they were quickly dismantled. Also huddled in Woodham's yard in October 1968 were 216 ex-BR steam locomotives, a 'Castle' class 4-6-0 being prominent on the right, and while all but four gradually departed over the ensuing years, destined for a multitude of preservation projects (Derby-built 4F 0-6-0 No. 43924 had left the previous month), regrettably no such favour was extended to the 'Warships'. No. D600 was dismembered in March 1970 while No. D601 remained intact until June 1980.

48

*Opposite above*: Having been diverted from its normal route through the Severn Tunnel, No. 1056 *Western Sultan* passes Tuffley Junction, Gloucester, with the 15.25 from Cardiff to Paddington on 16 September 1975. The train will travel via Stroud to Swindon, from where it will continue along its usual path to London. The lines diverging to the left, subject to a 50 mph speed restriction, led to Gloucester Eastgate station which closed on 1 December 1975, since when all trains serving Gloucester have had to use the former Central station.

*Opposite below*: On the same day at Horton Road Junction, Gloucester, No. 1001 *Western Pathfinder* heads a rake of Yeoman stone hoppers emanating from their Merehead quarry, near Cranmore in Somerset, towards South Wales. The train is approaching from the Cheltenham direction, while the line on the right, which the train in the previous picture will have followed after passing through Gloucester Central station, joins the Birmingham-Bristol main line at Gloucester Yard Junction. To the left on the far side of the level crossing was the entrance to the old Gloucester GWR shed, while opposite a cinder path led to the former Midland Railway Gloucester (Barnwood) shed. When released from Swindon Works in February 1962, No. 1001 was experimentally painted maroon, later adopted as standard for the class. Fourteen years and nine months later it was withdrawn in October 1976, as the longest-lived 'Western' in BR service.

*Right*: No. 1048 *Western Lady* nears Lansdown Junction as it runs through the Regency town of Cheltenham Spa, on the ex-Midland Railway route between Birmingham and Bristol, with empty china clay wagons returning from Stoke-on-Trent to Cornwall, again on 16 September 1975. A passenger train stands at the northbound platform. The rails lower right were once owned by the Great Western Railway and served Cheltenham Malvern Road and St James stations, before continuing to Stratford-upon-Avon and Birmingham. This was closed as a through route in 1976, but the heritage Gloucestershire Warwickshire Railway has since reopened in stages the section between Cheltenham Racecourse and Broadway. No. 1048 survived until the very end of the 'Western's' existence on BR, shadowing the 'Western Tribute' special along with No. 1010 *Western Campaigner* on that fateful 26 February 1977 (see inside back cover). Fortunately *Western Lady* was one of the lucky ones to escape the clutches of the scrap metal merchants, and today is cared for on the Midland Railway – Butterley, near Ripley, Derbyshire.

No. 1025 *Western Guardsman* pulls away from Stratford-upon-Avon, after just being attached to a Locomotive Club of Great Britain charter *en route* to Newport via Gloucester on 26 October 1974. Earlier the special had left Paddington behind No. 1069 *Western Vanguard*, but at Didcot this loco had given way to preserved 'Merchant Navy' class 4-6-2 No. 35028 *Clan Line* for the run to Shakespeare's birthplace. 'Guardsman' was no stranger to South Wales, having spent its early years based at Cardiff Canton shed before moving to Laira in April 1966, and then to Bristol Bath Road the following November. It returned to South Wales in January 1969 but this time to Landore before a final move back to Laira in October 1971. It was condemned in October 1975, although it was not broken up at Swindon until January 1979.

No. 1065 *Western Consort* calls at Banbury with a Paddington-Birmingham service on 4 October 1975. From Didcot the 'Western' had followed preserved 'Castle' class 4-6-0 No. 7029 *Clun Castle* with a returning steam special to Tyseley, Birmingham. The latter is taking on water before continuing its journey. Completed at Swindon under BR in May 1950, *Clun Castle* was the last of its class on the books of BR when withdrawn in December 1965. The younger *Western Consort* left Crewe Works new in June 1963, and was to be seen on services that would regularly have been in the hands of the 'Castles' in earlier years. Regrettably the sands of time were beginning to run out for No. 1065, which was deleted from stock in November 1976.

*Opposite*: An ambitious trip, 'The Western Talisman' organised by the Western Locomotive Association, took No. 1023 *Western Fusilier* down the East Coast main line from London King's Cross to York on 20 November 1976. No. 1023 was the last 'Western' to receive an overhaul at Swindon Works in September 1973, and had recently been repainted at Laira for its role on some of the many specials organised during the last months of the class on BR. Here *Western Fusilier* accelerates the ground-breaking excursion away from the 40 mph speed restriction over Selby swing-bridge, after crossing the River Ouse. Despite the cold a number of exuberant 'Western' enthusiasts can be seen at the open windows of the leading carriages, while rising above the third coach is the tower of Selby Abbey. At the insistence of the Eastern Region a black film with two translucent white discs (the domino effect) had been applied to the train indicator panels. This was the only 'Western' so modified.

*Below*: During the layover at the Minster city before returning to the capital, No. 1023 was stabled in the depot yard. It was open house and dozens of people, both young and old, took advantage of the opportunity to visit the cab. I don't recollect whether or not they managed to get the pushchair on board!

No. 1023 *Western Fusilier* was again the chosen power for the Railway Pictorial Publications' 'Western Memorial' railtour from Paddington, via Severn Tunnel Junction, Hereford and Shrewsbury, to Crewe on 29 January 1977. Note the Great Western Railway coat of arms and the BR crest either side of the headboard. Having travelled over former London & North Western Railway tracks from Shrewsbury, No. 1023 is just over one mile away from its destination as it passes the 18-lever Gresty Lane No. 2 signal box. The GWR did have running powers over this section of line from Nantwich. The overhead electric wires go no further south. This was an appropriate destination for one of the last specials, which on the return journey via Chester passed alongside Crewe Works, the birthplace of Nos. 1030 to 1073, three of which – Nos. 1041, 1048 and 1058 – survived into 1977. The other four that remained on the books of BR at the start of January 1977 – Nos. 1010, 1013, 1022 and 1023 – emanated from Swindon Works.

On a bitterly cold 12 February 1977, No. 1023 *Western Fusilier* again visited York, this time starting from Exeter St David's with the F&W Tours' 'Western Finale' excursion, which travelled via Birmingham, Derby and Chesterfield. Here threading the West Yorkshire coalfield, running some one-and-a-half hours late after striking a cow near Charfield, between Bristol and Gloucester, 'Fusilier' nears Goose Hill Junction, Normanton. This is ex-Midland Railway territory, although the tracks from the Wakefield direction to the right of the signal box by the rear carriage, were once owned by the Lancashire & Yorkshire Railway. The two routes converged at Goose Hill. The cooling towers of Wakefield Power Station can be discerned above the leading coach, while St John's Colliery can be seen at a higher level on the right.

Thirty-five minutes after the photograph overleaf was taken, after passing through Leeds and now on former North Eastern Railway tracks, the 'Western Finale' special speeds by Peckfield Colliery at Micklefield, on the last stage of its journey to York. Peckfield signal box controlled the access points to the colliery, which in the 1970s produced over 400,000 tons of coal per annum and employed some 500 men. During the layover period at York the locomotive was stabled outside the National Railway Museum, which was to become its home after retirement later that month.